Abhinav Mishra

Analysis of renewable energies in India

GRIN Publishing

GRIN - Your knowledge has value

Since its foundation in 1998, GRIN has specialized in publishing academic texts by students, college teachers and other academics as e-book and printed book. The website www.grin.com is an ideal platform for presenting term papers, final papers, scientific essays, dissertations and specialist books.

Visit us on the internet:

http://www.grin.com/

http://www.facebook.com/grincom

http://www.twitter.com/grin_com

Abstract

India is a developing country and it faces the problem of ever increasing energy demands. Fossil fuels are the primary source of energy but there aren't enough reserves of fossil fuels to meet the energy demands. Fossil fuels will ultimately run out and the mankind will be left with no option but to switch to alternative fuels.

Alternative sources of energy are non-conventional sources of energy i.e. other than fossil fuels. They are clean and never-ending sources. However, they initially require high investment costs and the technology to harness renewable sources of energy has barely hit puberty. Nevertheless, they have potential to overtake fossil fuels as primary sources of energy when there will be none to cater the needs of the masses.

The scope of this paper is to analyse the efficacy of non- conventional sources of energy. The object of this paper is to suggest credible alternatives to meet the energy demands of developing India.

Keywords- Non- conventional sources, fossil fuels, renewable resources.

Introduction

India is the third largest consumer of energy after China and USA. About 70% of country's electricity is generated through fossil fuels. Coal accounts for 40% country's energy followed by oil (24%) and natural gas (6%)[1]. Fossil fuels act as a double- edged sword. On one hand, they cause pollution and on the other hand, India has to spend bulk of its foreign currency in importing crude oil and natural gas. India is the fourth largest carbon emitting country in the world. Further, India's forest cover is lower than the global average. Global average forest cover is 30.8% of the total land area whereas India's forest cover is 23.8%[2]. This means that India's produces more pollution than its forest cover could absorb. Thus, there is an urgent need to switch to non-polluting alternative energy sources before it is too late.

Solar Energy

India has an advantage of being a tropical country and majority of landmass has nearly 300 clear sunny days. In theory, total solar energy falling on earth's surface is about 5000 trillion

---

[1] Eric Yep, India's Widening Energy Deficit, The Wall Street Journal, 9th May 2011. Available on http://blogs.wsj.com/indiarealtime/2011/03/09/indias-widening-energy-deficit/. Last visited on 14-02-2017.
[2] Forest area (% land area), The World Bank. Available on- http://data.worldbank.org/indicator/AG.LND.FRST.ZS. Last visited on 14-02-17.

kilowatt per year which is more than the entire output of available fossil fuels in the country.[3] Solar energy can be utilised in a number of ways. It may be utilised by converting it into electrical energy by photovoltaic cells or by converting it into heat energy for using in solar water heaters or solar cooker.

Advantages

- It is a clean and non-polluting source of energy.
- It is a never ending source.
- Can be used in remote locations which have not been electrified.
- Solar cells are long lasting and have low maintenance costs.

Limitations

- Solar energy cannot be produced during night.
- Energy production gets low when weather is cloudy.
- Solar cells initially require high investment.
- Solar cells are not very effective in production of electrical energy. The most effective photovoltaic cells convert not more 20% of the energy they receive.[4]

Suggestions

- Use of solar concentrators- A solar concentrator is a device through which sun rays can be concentrated over a specific area. By using solar concentrators, output can be increased by upto 50%.[5]
- Use of a newer, effective technology- Currently, traditional crystalline silicon cells are used for production of electricity which have an efficiency of about 20%. Scientists have already found a much more effective substitute to it. Multi junction gallium arsenal cells have achieved efficiencies upto 43.5%.[6]
- Use of tubular batteries instead of lead acid batteries- Lead acid batteries are the most commonly used batteries. They last about 3-4 years and require regular maintenance.

---

[3] 'Solar', Ministry of New and Renewable Energy, Government of India. Available on http://www.mnre.gov.in/schemes/grid-connected/solar/ Last visited 15-02-17.
[4] Solar Power, www.tc.unm.edu. Available on http://www.tc.umn.edu/~dama0023/solar.html Last Visited 15-02-17.
[5] http://www.electroschematics.com/8280/maximizing-solar-panel-efficiency-and-output-power/ Last visited on 15-02-2017.
[6] http://sinovoltaics.com/solar-cells/solar-cell-guide-part-3-third-generation-multi-junction-gallium-arsenide-gaas-solar-cells/ Last visited: 15-02-17.

On the other hand, tubular batteries last for about 7-8 years and are more efficient than lead acid batteries.[7] In this way, one of the major disadvantages of solar energy i.e. it cannot be produced at night, can be negated to a great extent.

Biomass

Biomass refers to organic matter derived from plants or other livings organisms. Since India is an agricultural country, it is available in abundance. Energy from biomass can be used either directly, by burning it or by converting it into more efficient forms such as biogas or bio-fuel. Majority of rural households use biomass to cater their energy needs. Some of the forms in which biomass can be used are discussed below.

Biogas

Use of biogas (commonly referred as gobar gas in India) is a sole technology which can completely overtake the use of Liquefied Petroleum Gas (LPG). It is produced by breakdown of organic material such as agricultural wastes, sewage, cowdung, etc, in the absence of oxygen. It chiefly contains methane which has high calorific value. It can be used for cooking or for making electricity from a heat engine.[8] Advance methods have shown that methane content can be increased upto 80-90% which is currently 40%.[9]

Conversion of biomass to biogas has an advantage in the sense that biogas produces very little carbon dioxide while burning and the remains can be used as manure.

Bio-diesel

Biomass can be used to produce biodiesel which can replace conventional fuels to a certain extent. Biomass can be converted to transportation fuels such as ethanol and biodiesel. It can be produced from crops such as sugarcane and corn.[10] There are several companies which collect used cooking oil from restaurants and convert it to biodiesel.[11]

Challenges

---

[7] https://www.okayapower.com/which-type-of-inverter-battery-should-you-choose/ Last visited: 16-02-17.
[8] "Biogas & Engines". , www.clarke-energy.com. Last visited: 22-02-17.
[9] Richards, B.; Herndon, F. G.; Jewell, W. J.; Cummings, R. J.; White, T. E. (1994). "In situ methane enrichment in methanogenic energy crop digesters". Biomass and Bioenergy. 6 (4): 275–274
[10] Energy Kids. Eia.doe.gov. Last visited: 22-02-17.
[11] "Types of Biofuels: Ethanol, Biodiesel, Biobutanol | Renewable Energy". Energy Digital. Last visited: 22-02-17.

- One of the major problems of biomass is that it is seasonal. For example, biomass will be abundantly available during harvesting season whereas there will shortage during off-season.
- It is less efficient as compared to conventional fuels.
- Biodiesel cannot be used in normal vehicles such as bikes, etc.
- Lack of technological know-how for harnessing biomass energy.

Suggestions

- Government should provide subsidies to those who are willing to set a gobar gas plant like provides subsidies to people installing solar water heaters, etc.
- Research and development should be encouraged in this field so as to increase its efficiency and reduce initial investment.
- Awareness programs should be organized in remote areas so that people are encouraged to switch to biogas.

Water

It may sound shocking but water can be a potential source of energy. Not just for production of hydroelectricity but also by using water directly as an element of production. Water is made up of hydrogen and oxygen and the "hydrogen" part can be used as a fuel. Hydrogen releases tremendous amount of energy per mole while burning. A major advantage of using hydrogen as a fuel is that it does not produce any kind of pollution. It produces water on burning. Many countries have claimed to have manufactured hydrogen power vehicles which run on water.

Challenges

- One of the major challenges is that more energy is consumed in separating hydrogen from oxygen than it is gained from burning hydrogen. Since water is a highly stable compound, a large amount is required to break water into its constituent elements. Thus, an energy deficit is created during the entire which makes unfeasible.
- It is very difficult to store hydrogen and since hydrogen is high inflammable, it has to be stored with great caution.

Suggestions

- Certain techniques for separating hydrogen from water have been developed which seem to be promising. For example, a scientific process called artificial photosynthesis has been developed for separating hydrogen from water by using sunlight. This process is inspired by photosynthesis which occurs naturally. In this process, an artificial leaf device is submerged in an aqueous solution which, when illuminated with a light source, forms hydrogen gas bubbles.[12]
- Further techniques can be developed in future if government encourages research and development in this field.

Waste

Waste-to-energy process been has been quite popular in a few developed countries for production of energy. In this process, energy is produced either in form of electricity or in form of heat. Some of the processes for producing energy from wastes are discussed below.

Incineration

Incineration is one of the processes where there is thermal treatment of waste. It involves combustion of waste to produce energy. This process is widely used in European countries for production of electricity. India also has a fully functional incineration plant at Delhi.[13]

Gasification

It is a process which involves treatment of waste material at high temperatures (upto 700 degree Celsius) with controlled amount of hydrogen and steam. The mixture formed after the completion of this process is a fuel.[14]

Advantages

- These processes can be used to avert the dangers of open dumping and land filling of wastes.

---

[12] https://www.sciencedaily.com/releases/2012/05/120523102057.htm Last visited: 24-02-17.
[13] http://timesofindia.indiatimes.com/home/environment/pollution/Delhis-waste-to-energy-plants-toxic-costly-inefficient/articleshow/46751552.cms Last visited : 25-02-17.
[14] Thermal Gasification of Biomass, International Energy Agency Task 33, www.gastechnology.org. Available at https://waste-management-world.com/index/display/article-display/368649/articles/waste-management-world/volume-10/issue-4/features/plasma-gasification-clean-renewable-fuel-through-vaporization-of-waste.html. Last visited : 25-02-17.

- Incineration of municipal waste avoids release of methane gas into the atmosphere.
- Volume of waste is reduced by almost 90% by combustion and can be easily disposed off.

Disadvantages

- Incineration causes a lot of air if the incineration plant is not equipped with sufficient gas purifiers and the fumes are released untreated. This has been the case in Delhi. Experts have said that the incineration plants in Delhi cause pollution and increase the toxicity of air.[15]
- Incineration plants take away the resources required for working of a recycling plant. In this way, they are again harming the environment since new products need to be manufactured every time which are already causing environmental problems.
- A good ratio of population lives on collecting waste and giving them to recycling plants. This part of population would lose livelihood.

Suggestions

- Gas filters relying on advanced technology should be imported from European countries and installed in incineration plants in India.
- Inspiration needs to be taken from Sweden where 7,00,000 tonnes of waste is imported to per year to supply its waste-to-energy facilities.[16]

Wind Power

Moving wind possesses kinetic energy which can be converted into electrical energy by a windmill. India has been blessed with a long coastline which are suitable for installation of wind energy farms. Coastal areas of Gujarat, Tamil Nadu, Andhra Pradesh, etc. are suitable for installation of wind energy farms. Hilly areas of western and eastern ghats and areas of Arunachal Pradesh are also suitable for production of electricity by tapping wind energy. These areas receive consistent winds of suitable velocity suitable for power generation.

Advantages

- Wind energy is free after the initial investment of installation of a windmill.

---

[15] Supra note 13.
[16] https://sweden.se/nature/the-swedish-recycling-revolution/ Last visited: 25-02-17.

- It is pollution and reliable and unlike solar energy, it can also be produced at night.

Limitations

- It requires high initial investment and the initial cost is greater than that of fossil fuels. Therefore, individuals don't install a windmill on their roofs.
- Wind farms are located far from the main city and dedicated transmission lines need to be laid down which adds to the initial cost.

Suggestions

- Wind farms need less area and don't affect vegetation on the ground. The utilise air space above land which is a national resource. It can be auctioned[17] the money can b utilised for funding research in the area of wind energy.
- Research has shown that initial cost of installing a windmill can be reduced. A cost installing a windmill on a rooftop can, now, be as low as buying a new iphone.[18] An Indian startup, Avante Garde Solutions, have manufactured a wind mill which is equal to the size of a ceiling fan and can generate 5 KWh/KW per day and requires intial investment of just $750.[19]

Nuclear Energy

Nuclear energy is obtained by breaking atoms of a heavy element such uranium, plutonium, etc. There are two process by which nuclear energy can be harnessed- fission and fusion. In the process of fission, a heavy atom is broken and large amount of energy is released which can be used to produce electricity. The other process is fusion in which two atoms are combined to produce energy. In this process, a larger amount of energy is released. However, the energy released in this process is uncontrolled and is only used to produce bombs.

Advantages

- Very less amount of fuel is required while producing nuclear energy.
- It is considered to be a renewable source since large reserves of uranium have been found which are almost inexhaustible.

---

[17] http://www.business-standard.com/article/pti-stories/govt-can-use-auctions-to-get-renewable-energy-projects-study-115051501529_1.html Last visited 28-02-17.
[18] https://e27.co/with-the-cost-of-an-iphone-you-can-now-buy-a-wind-turbine-that-can-bring-affordable-clean-energy-to-your-home-for-lifetime-20160606/ Last visited 28-02-17.
[19] Ibid

- The process of fusion produces clean energy and no waste is left behind for disposal.

Limitations

- There is always a danger of radioactive leakage.
- In the process of fission, radioactive waste is produced which is hard to dispose.
- The process of fusion cannot be controlled to produce limited energy.

Suggestions

- Recent researches have shown that fusion reactions can be controlled to produce energy.[20] Such techniques can be imported to produce clean nuclear energy since fusion does not create any waste and if the reaction is controlled, the chances of radioactive leakage are also nill.

Conclusion and suggestions

Since fossil fuels are depleting fast, India needs to search for reliable alternatives. Though the alternatives are already present, they have not been harnessed to their full potential. Since installation of a wind farm or a solar plant cannot be agenda to win elections, little heed is paid by the governments to promote these alternatives. A political will along with the support of general public is necessary for a complete switchover to these alternatives in near future.

Research in the field of finding alternatives should be at par with country's space research program to gain massive success.

---

[20] https://www.visionofearth.org/industry/fusion/how-do-we-turn-nuclear-fusion-energy-into-electricity/ Last Visited: 28-02-17.

YOUR KNOWLEDGE HAS VALUE

- We will publish your bachelor's and master's thesis, essays and papers

- Your own eBook and book - sold worldwide in all relevant shops

- Earn money with each sale

Upload your text at www.GRIN.com
and publish for free